The Role of Manganese as a Controller for Gold Mineralization in the Serpentinites of the San José de las Malezas Gold-Quartz Deposit in Santa Clara, Villa Clara, Cuba.

Ricardo A. Valls, P. Geo., M. Sc.

ISBN-13: 978-1533509000
ISBN-10: 153350900X

Contents

Summary

The San José de Las Malezas quartz-gold deposit is located on the North side of the Cuban Ophiolitic Complex, within the Structural-Facial Zone "Zaza", Province of Villa Clara, Cuba. In the area of the deposit there is a well-developed listwanite[1] zone that begins with (i) massive and relatively little altered serpentinites that pass into (ii) an oxidized serpentinites which are in gradational contact with (iii) a completely carbonated rocks and (iv) a core of silicified rocks and quartz veins.

Copper and gold mineralization are associated with these quartz bodies, but binary correlation studies of geochemical data suggest that the ores are genetically independent of one another (Valls and Cruz, 1995a).

In concordance with the proposed genetic model supported by thermodynamic calculations and field observations, copper is thought to be introduced by a hydrothermal process, whereas gold was leached from the serpentinites.

The evolution of manganese minerals from anaerobic to aerobic species provoked a reducing environment which hampered the precipitation of gold and other ores inside and near the serpentinites. Accordingly, a significant negative correlation between gold and

---

[1] Listwanite (also sometimes spelled listvenite, listvanite, or listwaenite) is unusual rock type that forms when ultramafic rocks (most commonly mantle peridotites) are completely carbonated (https://en.wikipedia.org/ wiki/ Listwanite).

manganese in similar areas of hydrothermally altered serpentinites, may be good indicators of possible gold concentrations.

The mechanical transportation of very small and thin scales of native gold from the serpentinites by the fluids is also suggested as a possible mechanism to explain the presence of this kind of scales inside the iron-altered zone.

Introduction

The San José de Las Malezas deposit is located within the Structural-Facial Zone (S.F.Z.) "Zaza", in the province of Villa Clara in Central Cuba. This S.F.Z. is composed of (i) a volcano-sedimentary complex of Lower Cretaceous age (Turonian) located to the South, (ii) the Ochoa Formation composed of limestones and marls of the Eocene in discordant contact to the North, and (iii) the Zurrapandilla Complex, composed of diabase porphyries, spilites, gabbro diabase, gabbro diabase porphyries, and other gabbroic rocks that cut both described complexes (Cabrera and Tolkunov, 1979).

Within the S.F.Z. "Zaza" and in tectonic contact with the volcano-sedimentary complex are serpentinitic bodies, that form a large massif with an east-west orientation. These intrusions form part of the Cuban Hyperbasitic Belt of the Upper Cretaceous, and they are commonly interpreted as the remains of an ancient oceanic crust.

These serpentinites are massive, fractured, light green rocks, with a reticular structure due to the uneven distribution of chrysotile and antigorite, and the presence of metallic minerals, such as magnetite, chromite and spinel. This mineralogical composition, suggests that the original rock was a harzburgite, however the original texture has been completely erased. The serpentinites are frequently cut by diabase, micro diabase and porphyritic diabase dikes from the Zurrapandilla Complex.

A more detailed geological description of the region and the outcrop, including maps and cross-sections, appears elsewhere (Valls and Cruz, 1995a).

In this paper we will study the formation and evolution of manganese minerals within these serpentinites and their relationship with gold mineralization.

Regional Geology

The San José de Las Malezas quartz-gold deposit is located within the Structural-Facial Zone (S.F.Z.) "Zaza", in the province of Villa Clara in Central Cuba (Fig. 1).

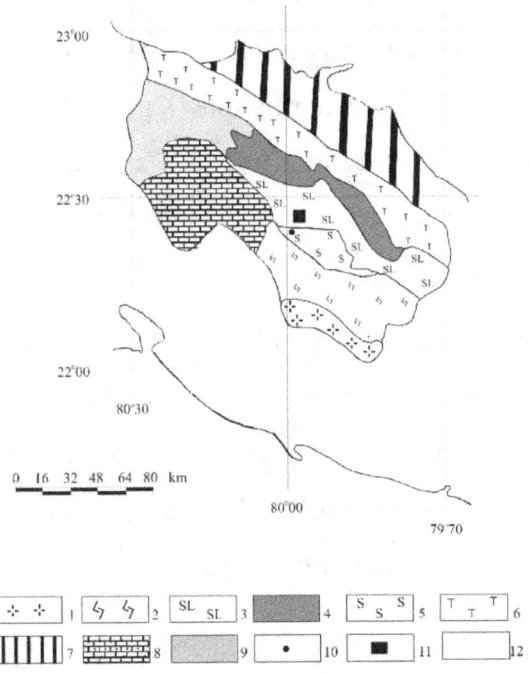

Figure 1. Regional geology of Central Cuba. 1.- Diorites; 2.- Volcanic sequences; 3.- S.F.Z. "Zaza"; 4.- S.F.Z. "Placetas"; 5.- Oceanic crust; 6.- S.F.Z. "Camajuaní"; 7.- S.F.Z. "Remedios"; 8.- Sedimentary formations; 9.- N-Q formations; 10.- City of Santa Clara; 11.- San José de Las Malezas deposit; 12.- Other formations (chiefly metamorphic sequences).

This S.F.Z. is composed of (i) a volcano-sedimentary complex of Lower Cretaceous age (Turonian) located to the south, (ii) the

7

Ochoa Formation composed of limestones and marls of the Eocene in discordant contact to the north, and (iii) the Zurrapandilla Suite, composed of diabasic porphyries, spilites, gabbro diabase, gabbro diabase porphyries, and other gabbroic rocks that cut both described complexes (Cabrera and Tolkunov, 1979).

As representatives of the S.F.Z. "Placetas" we have the Constancia Formation (J_3t), the Veloz Formation (J_3 t-K_1b), the Fidencia Formation (K_1b), the Carmita Formation (K_2al-cm), and the Vega Alta Formation (Palaeocene). They are mainly carbonates, with minor conglomerates, sandstones and also siliceous minerals.

Within the S.F.Z. "Zaza" and in tectonic contact with the volcano-sedimentary complex just described, we also find serpentinitic bodies, that form a large massif with an east-west orientation. These intrusions form part of the Cuban hyperbasitic belt of the Upper Cretaceous, and they are commonly interpreted as the remains of an ancient oceanic crust.

The serpentinites are massive, fractured, light green rocks, with a reticular structure due to the uneven distribution of chrysolite and antigorite in the rock, and the presence of magnetite, chromite and spinel. They are frequently cut by diabase, micro diabase and porphyritic diabase dikes from the Zurrapandilla Suite.

GEOLOGY OF THE DEPOSIT

In the vicinity of this deposit we find serpentinites, gabbroic rocks, diabase dikes, diorites, and quartz veins (Fig. 2).

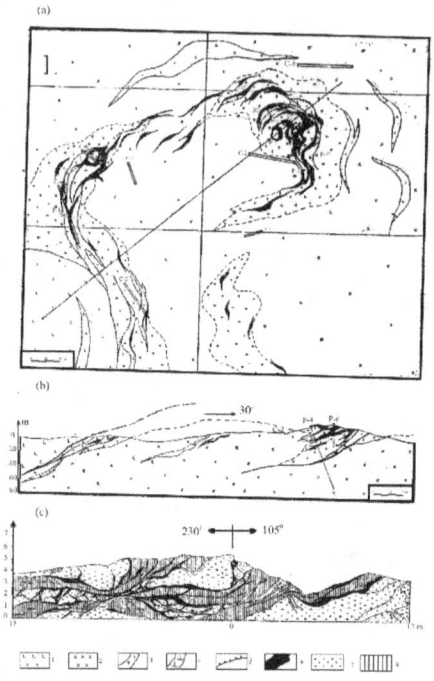

Figure 2. Geology of the San José de Las Malezas quartz-gold deposit, in Santa Clara, Cuba (From Cabrera and Talkunov, 1979). (a)- Local geology; (b)- Geological cut; (c)- Wall of the open pit. 1.- Serpentinites; 2.- Diorite porphyry; 3.- Dike of porphiritic diorites; 4.- Dike of porphyritic diabases; 5.- Tectonic faults; 6.- Ore bodies; 7.- Altered rocks; 8.- Quartz veins.

Serpentinites crop out to the west. At the centre of the area we find a big intrusion of gabbros, represented by grey, fine to medium grained rocks with a gabbroic texture. Both the gabbros and the serpentinites are cut by diabase, micro diabase and porphyritic diabase dikes.

These are dark grey compact rocks, fine grained to microgranular, with ophitic and microphytic structures.

We also find quartz veins that vary from 10 to 100 meters in length and 0,2 to 5 meters in width. The morphology of these quartz bodies is very complicated. Most frequently they form small lens, with abundant apophasis, wedges, and stockwork-like structures, intercrossing with subordinate diagonal veins. They also form sinusoidal structures like "waves". Post-ore tectonic movements added more complications to the structure of these formations. The more abundant quartz is a milky white variety, that frequently turns reddish because of the presence of hematite.

Sometimes one can observe the presence of porous quartz, due to the weathering, with the pores filled with limonite and other iron minerals. Copper mineralization occurs where the quartz veins are in contact with the fractured host rocks, and forms pockets, nests, crusts, and impregnations of altered minerals, containing an uneven distribution of gold.

The Formation of the San José de las Malezas Deposit

Our model begins with the protrusion of an ultramafic body of harzburgite composition through a ridge axis. It has been suggested elsewhere (Valls and Gonzalez, 1987), that this ultramafic magma could have been subjected to a partial melting process, due to which we could have obtained an area of gold enrichment by gravitational separation inside a magma chamber.

After the protrusion, the sea water and the heat from the upwelling zone initiated the serpentinization of the rocks. During this process, Mn^{2+} was liberated because of the decomposition of olivine. Another possible mineral that could liberate Mn^{2+} during the serpentinization of these rocks is pyrophenite ($MnTiO_3$) (W. Trzcienski, personal communication).

Under these anaerobic conditions, only divalent manganese minerals could be formed. Two possible contributing reactions are given bellow, (i) the formation of pyrochroite from tephroite (1), and (ii) the formation of rhodochrosite from tephroite by carbonatic sea water (2).

(1) $Mn_2SiO_4 + 2H_2O ---> 2Mn(OH)_2 + SiO_2$

(2) $Mn_2SiO_4 + 2H_2CO_3 ---> 2MnCO_3 + SiO_2 + 2H_2O$

Since pyrochroite is a much less common mineral than rhodochrosite (Crerar et al., 1976), we can assume that the formation of rhodochrosite was the most probable reaction. In fact, when we study the mineralogy

12

of this type of deposit World-wide, we usually find references to the presence of rhodochrosite, rhodonite, pyrolusite, and other manganese minerals (Baranova and Ryzhenko, 1981; Farfel, 1984; Baranova and Koltsov, 1987; Camus, 1990; Rodriguez and Warden, 1993, etc.).

The formation of rhodochrosite will take place for as long as olivine is decomposed during the serpentinization of the rocks in the heated zone near the rift.   Since we find relics of olivine crystals in these serpentinites, we can assume that serpentinization was stopped before the obduction onto the colliding continental plate during the final stage of the ocean closure.

After the obduction of these rocks onto an aerobic environment, the evolution of the manganese minerals responded to the increasingly oxidizing conditions, as shown in Fig. 3. First we have the oxidation of the rhodochrosite into hausmannite (3), second the oxidation of hausmannite into bixbyite (4), third the hydration of bixbite into manganite (5), and finally the oxidation of manganite into pyrolusite (6).

(3) $\qquad 3MnCO_3 + \frac{1}{2}O_2(g) \longrightarrow Mn_3O_4 + 3CO_2(g)$

(4) $\qquad Mn_3O_4 + \frac{1}{2}O_2(g) \longrightarrow 3Mn_2O_3$

(5) $\qquad Mn_2O_3 + OH^- + H^+ \longrightarrow 2MnOOH$

(6) $\qquad 2MnOOH + \frac{1}{2}O_2(g) \longrightarrow 2MnO_2 + H_2O$

Obduction also provoked the crushing of the rocks and the development of several tectonic systems, the main of which had a NE orientation.   Along these fractures we observe dikes of diabase from the Zurrapandilla Complex. During the intrusion of these dikes, the area was affected by a low temperature hydrothermal-metasomatic process. This process developed a well formed listwaenitic zone to which the mineralization is related. The provenance of these fluids is yet to be established, but we can propose three possible origins:

a.- Magmatic origin - orthomagmatic fluids related to the Zurrapandilla magmatic complex.

b.- Slab origin - sea water fluids related to the dehydration zone of the subducted slab.

c.- Mixed origin - fluids that are the result of a combination of the first two options.

Studies done by Ploshko (1963), Zuffardi (1977), Pipino (1980), Buisson and Leblanc (1986), and Pallister et al. (1987) on similar listwaenitic zones concluded that these fluids were composed mainly by H_4SiO_4, H_2CO_3, H_2O, H_2S, K, Na, Rb, and probably $CO_2(g)$ and $CH_4(g)$. Mottl (1991) suggests that these type if fluids usually present high values of pH, high carbonate alkalinity, and low chlorine. In accordance with the mineralogical associations in the area of San José de Las Malezas (Valls, 1995b), I believe that the main metallic component of this fluid should have been copper, with lesser amounts of lead, zinc, silver, and arsenic.

These fluids reactivated the serpentinization of the rocks, so more manganese was released from the remaining olivine crystals. Here, the most probable reaction should have been first- the formation of rhodonite (7), second the oxidation of rhodonite into hausmannite (8), and then the same evolution pattern as shown in equations (4 - 6) to arrive to the formation of pyrolusite.

(7) $\qquad$ $Mn_2SiO_4 + H_4SiO_4 \longrightarrow 2MnSiO_3 + 2H_2O$

(8) $\qquad$ $MnSiO_3 + \frac{1}{2}O_2(g) \longrightarrow Mn_3O_4 + SiO_2$

All these processes are schematically represented in Fig. 3. The formation of these manganese oxides leads to an increase in $fO2$ making difficult the precipitation of gold, copper and other ores in and near the serpentinitic bodies.

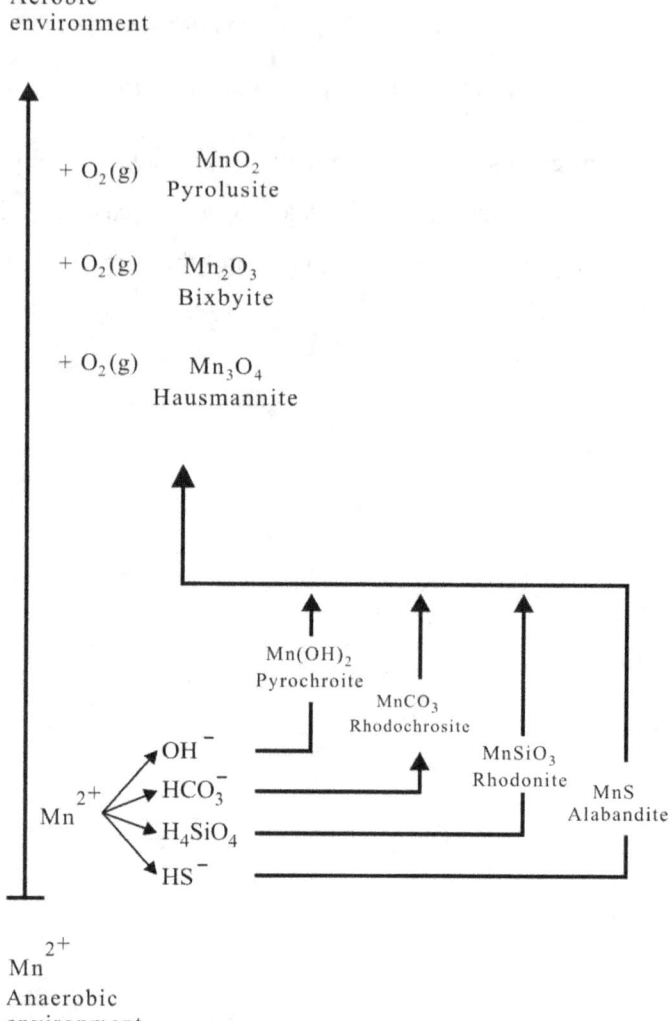

Figure 3. Evolution of Mn2+ in response to increasingly oxidizing conditions, near 283 °K, and 1 atm.

Dynamic of the Hydrothermal Fluid

Although it has been considered in the past as a typical copper-gold deposit, the correlation study of the existing data from San José de Las Malezas quartz gold vein deposit (Valls, 1995b) shows no significant correlation between gold and copper, and a strong positive correlation between gold and lead in the siliceous zone. Therefore, we conclude that copper and gold are spatially, but not genetically related.

In the proposed model, copper was contributed by the hydrothermal system, with Pb, Zn, As, and Ag, while gold was provided by the serpentinites hosting the quartz bodies. A schematic representation of this process is shown in Fig. 4. Since we already deal with the effect of reactivation of the serpentinization of the host rocks, we will focus on the mechanism of transportation and deposition of the ores.

The listwaenitic alteration consists of four zones, (i) a less altered serpentinites, (ii) an iron-altered zone, (iii) a carbonatic altered zone, and (iv) a siliceous zone (Sawkins, 1990, Valls, 1995b, et al.). These lenses grade laterally into the less altered serpentinites through a talc-carbonated zone.

According to the geochemical results of a meter by meter channel sampling through the four zones (Valls, 1995b), gold, copper, silver, zinc, arsenic and lead were found to concentrate mainly in the siliceous zone. Gold also was found concentrated inside the iron-altered zone, while the carbonate-rich zone is almost barren of ores.

Ricardo A. Valls

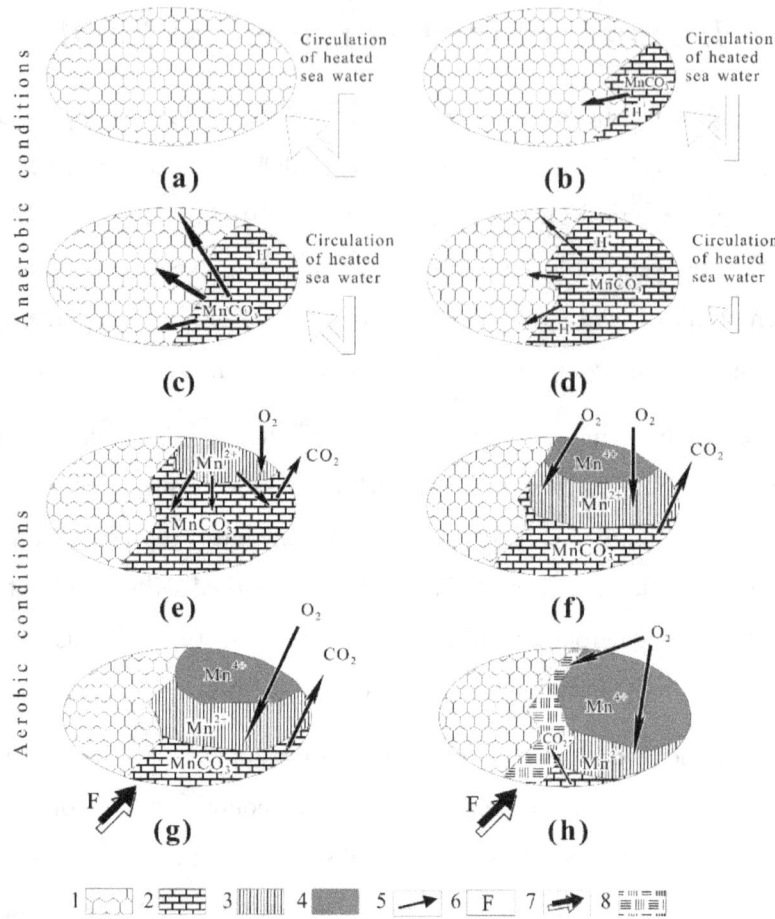

Figure 4. Schematic representation of the alteration of an ultramafic massif and the evolution of the manganese minerals from an anaerobic to an aerobic alteration after the obduction onto the surface of the massif. See text for detailed explanations. 1.- Ultramafic massif, 2.- Formation of rhodochrosite, 3.- Formation of hausmannite, 4.- Formation of manganite, byxbyite and pyrolusite, 5.- Vector of alteration, 6.- Ore fluid, 7.- Vector of mineralization, 8.- Listwaenitic zone.

It is commonly assumed that gold is transported in the (+1) oxidation state (McKibben et al., 1990). Since gold is a soft electron acceptor, it should form especially stable complex with soft ligands as HS^-.

A probable reaction is its transportation as a thio-complex (9).

(9) $Au + 2H_2S + \frac{1}{4}O_2(g) \dashrightarrow Au(HS)^-_2 + \frac{1}{2}H_2O + H^+$

I also believe that very small and thin scales of native gold could have been mechanically removed by the fluids from the serpentinites. The flat form of these grains allows them to be transported very easily, as seen from our experience during panning to obtain heavy concentrates from these and similar zones. This mechanism of transportation is doubtless less efficient than the one represented earlier (9), but it helps to explain the existence of this kind of native gold scales in the iron-altered zone (Fig. 5).

As the hydrothermal fluid moves toward the surface, several important factors will control its stability.

Inside and near the serpentinites, the evolution of the manganese minerals to their trivalent states, will consume oxygen provoking a reducing environment that difficult the precipitation of gold and other ores.

Closer to the surface, we have first the loss of temperature due to the mixing with meteoric waters, and second, the presence of a more oxidizing environment favoured by the existence of faults and fractures of the rocks.

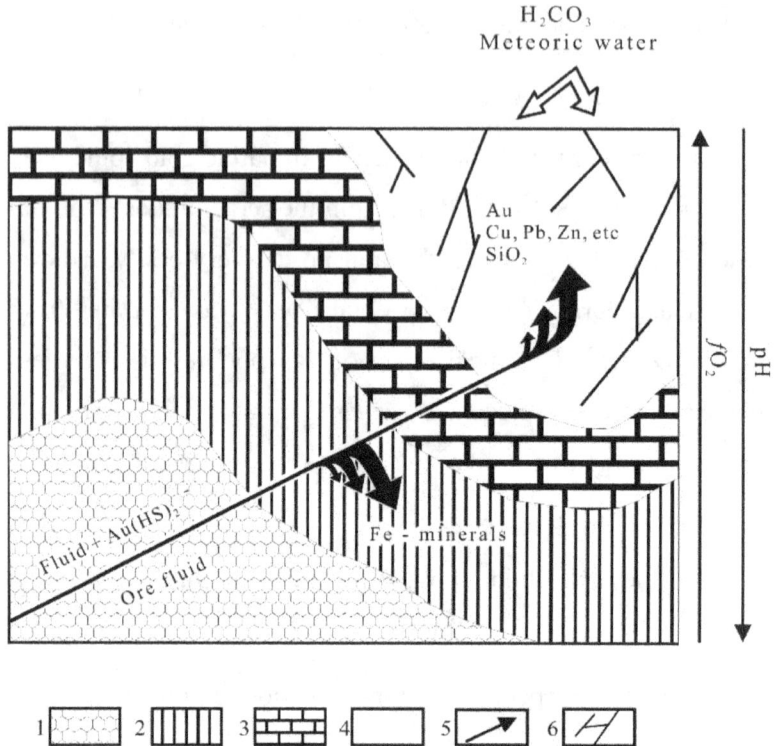

Figure 5. Dynamic of the ore fluid. This model proposes two mechanisms for gold transportation: (i) mechanically -as thin scales of native gold-, and (ii) as a thio-complex. Mechanical transportation explains the presence of scales of native gold in the iron altered zone. Main gold and copper concentration are in the siliceous zone, where the ore fluid got mixed with meteoric waters in a more acidic environment, with low values of fO₂ due to the formation of massicot or galena (see text for further details) 1.- Less altered serpentinites or unaltered ultramafic rocks, 2.- Iron-zone, 3.- Carbonate-zone, 4.- Siliceous zone.

Both the decreases of the pH and the increment of the fO_2, will provoke the precipitation of gold. The precipitation of Au because of a decrement of pH is shown in equation 10 (Spycher and Reed, 1989; McKibben et al., 1990). Equation 11 shows the precipitation of gold due

20

to an increment of the fO_2 (Chris Gammons, personal comunication).

(10) $8Au(HS)_2^- + 6H^+(aq) + 4H_2O(aq)$ ---> $8Au(s) + SO^{2-}_4(aq) + 15H_2S(aq)$

(11) $4Au(HS)_2^- + 15O_2(g) + 2H_2O(aq)$ ---> $4Au(s) + 8SO^{2-}_4(aq) + 12H^+$

Very often we find strong correlations between gold and lead and we find native gold in galena and lead oxides like massicot (PbO) and crocoite ($PbCrO_4$). This leads us to assume that another possible mechanism of gold precipitation is the formation of galena or massicot as it is represented in equations 12 and 13.

(12) $4Au(HS)_2^- + 4PbCl_2 + 7O_2(g) + 2H_2O(aq)$ --->

$4Au(s) + 4PbS + 8Cl^- + 4SO^{2-}_4(aq) + 12H^+$

(13) $4Au(HS)_2^- + 4PbCl_2 + 15O_2(g) + 6H_2O(aq)$ --->

$4Au(s) + 4PbO + 8Cl^- + 8SO^{2-}_4(aq) + 20H^+$

These reactions could explain the strong positive correlations between gold and lead in the siliceous zone (Valls, 1995b).

Ricardo A. Valls

Conclusions and Recommendations

Based on the information available up to this moment, a model has been presented to explain the geochemical characteristics of the ore distribution in the San José de Las Malezas gold-quartz vein deposit, in Santa Clara, Cuba.

This model describes the serpentinization of an ultramafic protrusion, which is later obducted onto the surface to form part of an ophiolitic complex. Special attention has been given to the formation and evolution of different manganese minerals during the serpentinization of the host rocks. The model helps to explain the negative correlation between gold and manganese over the serpentinites.

It has been suggested elsewhere the possibility of a partial melting process of this ultramafic body before its protrusion. This process could have provoked the formation of a gold enriched zone due to the gravitational separation of this mineral. Although the partial melting of these rocks is possible, this idea needs to be tested in the future. The existence of a gold enriched zone in the serpentinites will not only help to explain the remobilization of gold in the hydrothermal-metasomatic fluid, but also it could become a prospecting objective in the area.

After the obduction, the area was affected by hydrothermal-metasomatic fluids along the developed tectonic system. The origin of these fluids is yet to be determined, but due to the presence of H_2CO_3, H_4SiO_4, and H_2S, it is possible to suggest a magmatic or magmatic-slab origin in preference to a pure slab source.

22

These fluids reactivated the serpentinization of the host rocks, and provoked the formation of a listwaenitic zone toward the less altered serpentinites.

The model assumes that gold was leached from the serpentinites by the fluids and considers two mechanisms for its transportation. The first one is the mechanical transportation of thin scales of native gold by the fluid. This may explain the existence of a gold enrichment zone in the iron-altered serpentinites.

The second mechanism is the transportation of gold as thio-complexes and its precipitation in the siliceous zone due to a decrease in pH and/or an increment of the fO_2 provoked by the formation of galena, massicot or other lead minerals (crocoite?), and the loss of temperature due to the mixing of the fluids with meteoric waters.

The same conditions of pH, fO_2, and loss of temperature in the siliceous zone, provoked the precipitation of copper, lead, zinc, arsenic and silver from the fluid, and the formation of copper and lead-zinc minerals in the contact of the quartz bodies with the crushed host rocks.

Although this model can presently explain all the known characteristics of the distribution of gold and other ores in this deposit, I am not presenting it as uncontroversial model, but as a working hypothesis to formulate what needs to be tested in future studies.

Acknowledgment

The initial sampling and analytical programmes were carried out by the author and his colleague Eng. Jorge Cruz in 1986, during the preparation of a project for the detail survey of the San José de Las Malezas deposit for the Geological Enterprise "Santa Clara". The Provincial Laboratory associated to the Geological Enterprise provided the analysis of the samples, including the determination of fluorine by a technique designed by the Bulgarian specialist Lazarina Lazarova Ignátova.

I wish to thank Dr. W. Trzcienski for his suggestions and discussions during the preparation of this paper, and also to Dr. Christopher Brooks and Dr. Alex Brown for their review and criticism. I want also to thank Dr. Chris Gammons in particular for his support and guidance, and also for his fair and constructive criticism. Finally, from the big community of the INTERNET, I specially thank Margaret Donelick of the O.D.P. Project and to Dr. P.J. Kenney from the British Geological Survey for their comments and suggestions on the nature and composition of the slabs fluids.

References

Baranova, N. N. and A. B. Koltsov. (1987). The influence of metals and volatiles in hydrothermal solutions on gold transport and deposition based on fluid-inclusions studies. Geochemistry International 24(1): 1.

Baranova, N. N. and B. N. Ryzhenko. (1981). Computer simulation of the Au-Cl-S-Na-H2O system in relation to the transport and deposition of gold in hydrothermal processes. Geochemistry International 18(4): 46.

Buisson, G. and M. Leblanc. (1986). Gold-bearing listwaenites (carbonatized ultramafic rocks) from ophiolite complexes. Metallogeny of basic and ultrabasic rocks. Inst. Min. Metall. 121.

Cabrera, R. and L. A. Tolkunov. (1979). Tipos y condiciones geolgicas de la localizacin de los yacimientos de oro en la zona septentrional de la antigua Provincia Las Villas. Ciencias de La Tierra y del Espacio. (1):

Camus, F. (1990). The geology of hydrothermal gold deposits in Chile. Journal of Geochemical Exploration. 36((1-3)): 197.

Clarke, F. W. (1924). The data of geochemistry. USA, Bulletin U.S. Geological Survey.

Crerar, D. A., R. K. Cormick, et al. (1976). Geochemistry of manganese: an overview. In Geology and geochemistry of manganese. 2nd International Symposium on Geology and Geochemistry of manganese., Sidney, Australia,

Farfell, L. S., N. I. Savelyeva, et al. (1984). Hydrothermal solutions at the Asku gold ore deposit. Geochemistry International. 21(1): 71.

McKibben, M. A., A. E. Williams, et al. (1990). Solubility and transport of platinum-group elements and Au in saline hydrothermal fluids: constrains from geothermal brine data. Economic Geology 85(8): 1926.

Mottl, M.J. (1991). Pore waters from serpentinite seamounts in the Mariana and Izu-bonin forearcs, leg 125: Evidence for volatiles from the subducting slab, in Proceedings of the Ocean Drilling Program. Scientific results. V 125, 373-385.

Pallister, J.S., J.S. Stacey et al. (1987). Arabian shield ophiolites and late Proterozoic microplate accretion. Geology (15): 320.

Pipino, G. (1979). Gold in Liqurian Ophiolites (Italy). International Ophiolite Symposium, Cyprus,.

Ploshko, V. V. (1963). Listwaenitization and carbonation at terminal stages of Urushten Igneous Complex, North Caucasus. International Geological Review (7): 446.

Rodriguez, C. and A. J. Warden. (1993). Overview of some Colombian gold deposits and their development potential. Mineralium Deposita 28(1): 47.

Sawkins, F. J. (1990). Metal Deposita in relation to plate tectonic. New York, Springer-Verlag, Berlin-Heidelberg.

Spycher, N. F. and M. H. Reed. (1989). Evolution of a Broadlands type Epithermal ore fluid along alternative P-T paths. Implications for the transport and deposition of base, precious and volatile metals. Economic Geology 84(2): 328.

Valls, R. A. and I. Gonzalez. (1987). Evaluación geomatemática de los muestreos litogeoquímicos en mina Descanso, Villa Clara, Cuba. Moa's Mining Metallurgical Institute.

Valls, R. A., J. Cruz Martin (1995a). Geochemical peculiarities of the Quartz-Gold Vein Deposit of San José de Las Malezas, Santa Clara, Villa Clara, Cuba. Université de Montréal, report.

Valls, R. A. (1995b). Geochemographic study of the mineralogical assemblages from the San José de Las Malezas Gold-Lead-Quartz segregational deposit in Santa Clara, Villa Clara, Cuba, Université de Montréal, report.

Zuffardi, P. (1977). Ore/mineral deposits related to the Mesozoic ophiolites in Italy. New York, Springer-Verlag, Berlin-Heidelberg.

About the Author

As a professional geologist with thirty-two years in the mining industry, I have extensive geological, geochemical, and mining experience, managerial skills, and a solid background in research techniques, and training of technical personnel. I am fluent in English, French, Spanish, and Russian. I have been involved in various projects world-wide (Canada, Africa, Russia, Indonesia, the Caribbean and Central and South America). Projects included from regional reconnaissance to local mapping, diamond drilling and RC-drilling programs, open pit and underground mapping and sampling, geochemical sampling and interpretation, and several exploration techniques pertaining to the search for diamonds, PGM, gold, nickel, silver, base metals, industrial minerals, oil & gas, and other magmatic, hydrothermal, porphyritic, VMS and SEDEX ore deposits. Special strengths are related to acquisition of new properties, geochemical and geological studies, management and organization, geomathematical analysis and modelling, compositional data analysis, structural studies, database design, QA&QC studies, exploration studies and writing technical reports. P.Geo. registered in the province of Ontario.